MW01626164

TRAVEL AND SLEEP
IN
PULLMAN
SAFETY AND COMFORT

New and Improved

PULLMAN ACCOMMODATIONS

IN THIS BOOK *New and Improved* Pullman accommodations are described in detail that help make going Pullman more than ever the safest, most comfortable way of going places fast.

Many of the *improvements* have already been made. Some of the *new* accommodations are generally available, too, while others, available now on a limited scale, await only the completion of new cars.

These *new and improved* accommodations are, we sincerely believe, the ultimate expression of comfort and safety while traveling. *Everything is there* that the mastery of travel hospitality can provide to help make your Pullman trip an adventure in good living.

In addition to the physical comforts and conveniences described, the courteous attention of Pullman employes proud of their art in serving you is always there.

Day or night, you enjoy *service, comfort* and *safety* in Pullman-operated cars, that no other way of going places fast can match!

So, pick the type of accommodation you prefer on the following pages — and *let Pullman be your travel host.*

To make each trip a pleasure every mile of the way

Here are the accommodations and services available to you when you go Pullman.

Some are probably familiar, *others* are entirely new, and *all* are designed to give you the ultimate in modern comfort while traveling.

Here you have the widest choice of travel accommodations that can be found anywhere — with everything you need to make your trip an adventure in gracious living enjoying traditional Pullman hospitality.

And in addition to the physical conveniences described, you are assured of those personal services that have been inherent in Pullman hospitality for more than eighty years.

Day or night, you enjoy *comfort, service,* and *safety* in Pullman-operated cars that no other way of going places fast can match.

So choose the type of accommodation you prefer and enjoy traveling *The Pullman Way.*

personal services . . . conveniences

Your Pullman car is temperature-controlled and air-conditioned. You may regulate these yourself if you have a private room.

Your accommodations are kept immaculate by Pullman's expert housekeepers.

Your porter will gladly assist you in connection with any telegrams you may wish to send.

There's always a "first-aid" kit in your Pullman car, available for immediate use.

If you want an extra blanket or pillow, the porter will be glad to oblige.

Want your shoes shined while you sleep? Just leave them under your berth or in the special shoe-compartment of your private room. The porter will do the rest.

If you desire a table to be set up in your accommodation, just ring for the porter.

You'll get the kind of a "brush-off" you like, from your courteous porter, just before you leave the train.

Whenever you want to get into or out of your upper berth, ring for the porter and he'll come with the ladder.

Like a paper bag to protect your hat? The porter will be glad to get you one.

The porter always uses fresh, spotlessly clean linen when making up your bed.

Just ring the bell, at your seat or in your room, whenever you want the porter.

You don't need to rumple your clothes on a Pullman. Every accommodation includes clothes hangers.

Every Pullman car has a water-cooler and sanitary drinking cups in a handy dispenser.

Pullman travelers may park pets in baggage car or in room accommodation if carried in suitable container.

For your convenience, every train with Pullman equipment has an up-to-date Hotel Red Book.

If you want to be awakened at any certain time, just tell the porter before you retire.

You don't have to burden yourself with heavy luggage on a Pullman. It's safe and sound in the baggage car.

Section

DAY

The section is the combination of the upper and lower berths containing all of the facilities of both berths. The upper berth provides adequate accommodations, with convenient lights, a shelf to hold toilet articles, hangers for your clothes, and bell with which to summon the porter at any hour. A safety ladder makes access and exit easy. The lower berth is furnished with individual lights, a hammock for clothes, shelf for sundry articles, two convenient clothes hangers, and bell to call the porter. Shoes placed under the lower berth will be shined by the porter. Roomy paper bags are available to safeguard your hat. Pillows will be

NIGHT

gladly furnished for daytime lounging. For passengers using section space, the porter will install a table for reading, food service, or games, if desired. Roomy washrooms are provided at either end of this Pullman car. Whatever the accommodations, Pullman attempts to bring to every traveler the comforts and conveniences to which he is accustomed at home. Here is a better travel life for everyone at a cost saving that appeals to economy-minded travelers. One need have no hesitancy about the sleep-inviting comforts of these accommodations — upper, lower, or section units of space provide the acme of travel comfort.

Duplex-Roomette

DAY

In its continuing search for greater travel comfort at lowest possible cost, The Pullman Company introduced the Duplex-Roomette car. The rooms, 24 in all, are arranged on each side of a center aisle with the lower room at aisle level and the upper room reached by two steps. In each room a seat with a sponge rubber back and seat cushion and the latest body-fitting contour is adjacent to a large window. Every Duplex-Roomette is equipped with complete toilet facilities, a comfortable full-length bed, individual control of heat, light and air-conditioning, cool drinking water, box for shoes to be shined by porter, ash tray, electric shaver plug-in

connection and plenty of coat hooks. Modern light facilities provide for general illumination, mirror light, and reading lights for seat and bed. Your porter will gladly install a table for writing letters or playing games. Liberal baggage storage space is provided in each room. The bed in the lower room, when not in use, slides under the floor of the upper room, and is easily pulled out over the seat for night use. In the upper room the bed is of the fold-in-the-wall type. When the passenger is ready to retire he lowers the folding bed, and returns it to its niche in the wall on arising. Both rooms have full solid sliding doors.

Roomette TYPE A

DAY

The "Roomette" is a completely enclosed, private room containing a pre-made bed, affording accommodation *for one person*. In daytime the bed folds flush into the wall at one end of the room, and the passenger has a sofa seat of the latest and most comfortable contour, with ample space for lounging, or for undressing before the bed is lowered for the nighttime arrangement. For dressing, the passenger can make the whole room space and its complete toilet facilities available by returning the bed to its niche in the wall. When the bed is made down for the night it is fastened at the foot by a lock. This is readily released when the passenger desires to

NIGHT

raise the bed upon arising in the morning. The door of the "Roomette" can be locked at night, or left open and a curtain drawn across the opening. The patron has many conveniences, such as individual regulation of ventilation, air-conditioning, heat and light; complete toilet facilities, with washstand folding into one wall, and above it a mirrored cabinet for toilet articles; a locker in which to hang clothes; a large shelf for luggage; cool drinking water, and a box from which the porter can remove shoes without disturbing the occupant. Ceiling, mirror and reading lights of new design provide ample illumination.

Roomette TYPE B

DAY

This Roomette, designed for single occupancy, is arranged lengthwise in the car, and utilizes every square inch of floor space to deliver the utmost in comfort and convenience. The comfortable bed, which comes out of the wall, is the folding type and is pre-made — all ready to get into. The toilet and folding washstand are in one corner of the room. There are lights for general illumination, and others for individual reading. There is a full-length mirror on the inside of the entrance door, and plenty of room for clothes in the wardrobe locker. The Roomette is fully air-conditioned, and there are separate thermostatic heat controls which

Roomette TYPE B

NIGHT

enable the passenger to regulate the temperature to suit individual preference. The circulating drinking water is mechanically cooled. At one end of the room there is a large luggage rack, and additional space for luggage beneath the seat. Additional conveniences include an electric razor outlet and folding armrest on the seat. The door can be locked at night or left open, if you prefer, and a curtain with a zipper drawn across the opening. There is ample space for lounging or undressing before the bed is lowered. This new Roomette is designed for the passenger who wants the advantages of a completely private room at low cost.

Duplex Single-Rooms

DAY

These private rooms, designed for single occupancy, are arranged crosswise in the car. They are available either level with the floor or three steps above it. Each room has a comfortable contoured sofa with a back that folds down to form a full-size bed. At one end of the room is a folding type toilet and a fixed wash basin.

As in other private Pullman rooms, these Duplex Single-Rooms accommodate a table for writing, food, or games. Lights are placed conveniently for both general illumination and individual reading, and there is an outlet for an electric razor. Each room is fully air-conditioned and has individual thermostatic heat controls which enable

Duplex Single-Rooms

NIGHT

the passenger to regulate the temperature according to individual preference. Other conveniences include a full-length mirror for dressing, and circulating drinking water that is mechanically cooled. There is ample luggage space, on the luggage rack and underneath the sofa. There are folding arm-rests on the sofa and a shoe locker in which shoes may be placed to be shined by the porter. And there's a connecting door to the adjoining room. This modern accommodation is a private sitting room by day and private bedroom by night. Pullman is making more of these Duplex Single-Rooms available to those who "go Pullman."

Connecting Double Bedrooms TYPE A

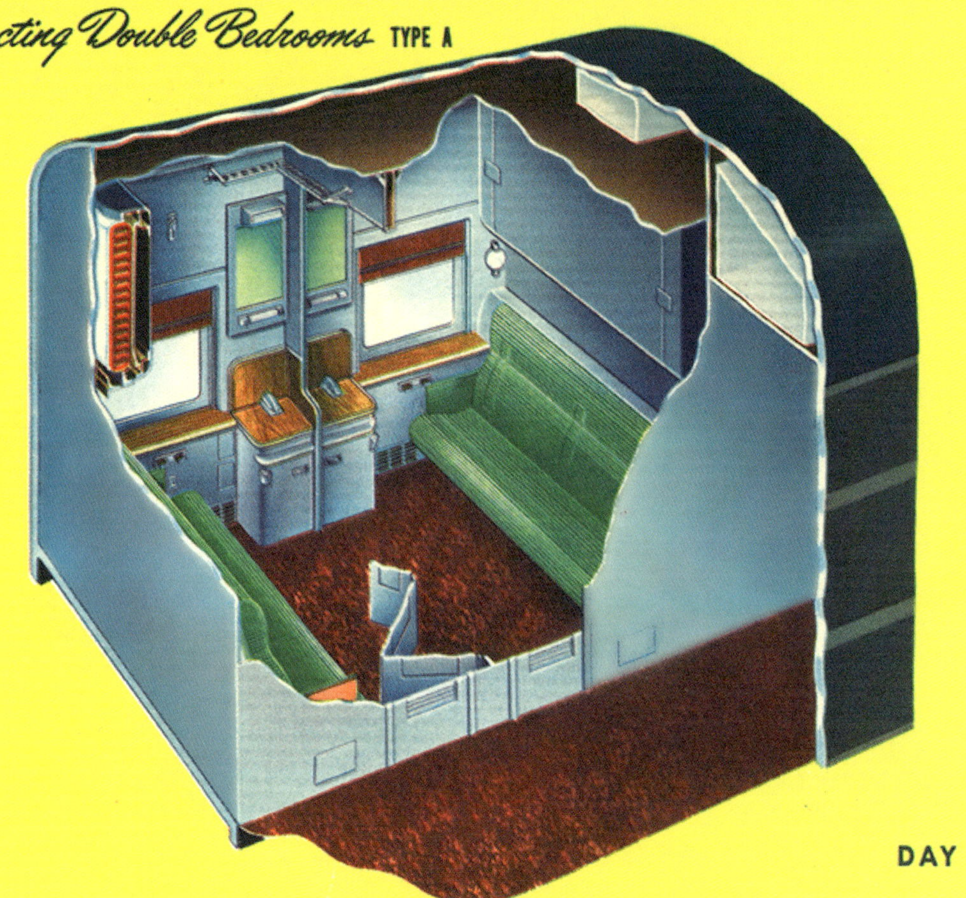

DAY

Each Double Bedroom is an enclosed private room, and contains a wide, comfortable sofa. For night time use this sofa forms a sleep-inviting bed. Above the sofa is an upper berth. The bedroom has a toilet which folds into a cabinet when not in use; also a washbasin. A table for writing and games or a service table for meals is provided. The room is provided with convenient lights, fan and separate heat control, that allows you to regulate room temperature to suit yourself. Each room is air-conditioned and includes cool, circulating drinking water, clothes hooks, and a full-length mirror. There is ample room for luggage space

Connecting Double Bedrooms TYPE A

NIGHT

beneath the sofa and in a luggage rack. Each room is provided with a locker for clothes. If you're seeking combined advantages of privacy, comfort and convenience the Double Bedroom is the accommodation. Some of the bedrooms are separated by a folding partition that permits the rooms to be kept as individual units, or they can be opened to make one large room where a table and folding chairs can be set for a meal or a game of cards to be enjoyed in complete comfort and privacy. Family groups find this a convenient and comfortable way to travel. Children may nap in one of the rooms without being disturbed.

Double Bedrooms TYPE B

DAY

These Double Bedrooms, which are crosswise in the car, add many comforts and conveniences to modern Pullman travel. Each has two beds. The lower bed is of the folding, pre-made, type and can be operated by the passenger. The upper berth, which is stored in the ceiling during the day, is operated by the porter. This new accommodation has two movable chairs which can be placed anywhere during the day. At night, one can be folded under the bed. One of the most attractive features of this Double Bedroom is an enclosed toilet containing also a folding washstand. Another convenience is provision for use of section tables for writ-

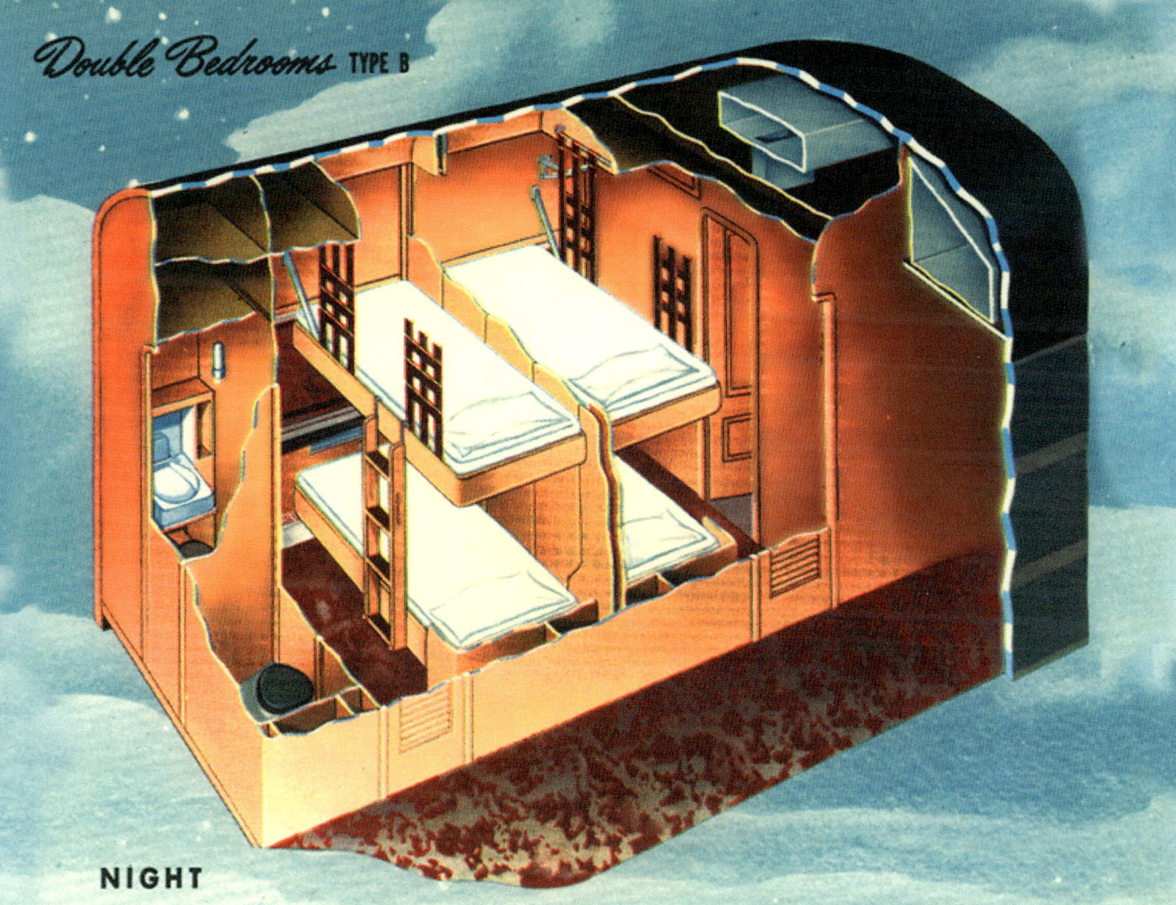

NIGHT

ing and games. The room has lights for both general illumination and individual reading, also a convenient electric razor outlet. Other conveniences include air-conditioning, individual thermostatic heat control, circulating drinking water that is mechanically cooled, a wardrobe locker, full-length mirror, a luggage rack, and a shoe locker. It is an ideal accommodation for two people. And the "economy" angle appeals to many husbands and wives, for this room costs no more than two lower berths! . . And if four people wish to travel together, the folding partition between the two bedrooms slides back to accommodate all.

Connecting Double Bedrooms TYPE C

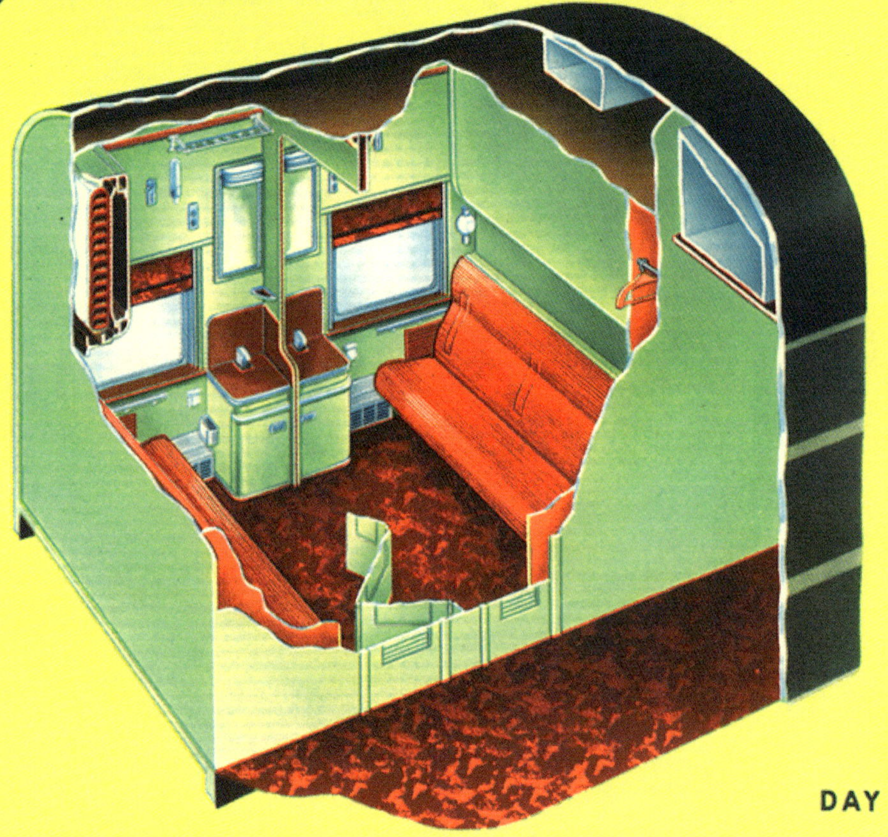

DAY

Connecting Double Bedrooms, each one designed to accommodate two people, are ideal for the accommodation of four people when the sliding partition between them is folded back. Each room has a comfortable, full-length sofa which converts into a bed at night. The upper berth is above the sofa. Each has a stationary washstand and a toilet which easily folds into a convenient cabinet. As in other bedrooms, there is provision for a service table. There are lights for general lighting, as well as individual reading lights. Each room is air-conditioned and the heat may be controlled to suit the preference of the occupants.

Connecting Double Bedrooms TYPE C

NIGHT

Additional conveniences include circulating drinking water that is mechanically cooled, a full-length mirror, wardrobe locker, a luggage rack, and additional luggage space beneath the sofa. There is also a locker for shoes which the passenger may wish the porter to shine. These connecting double bedrooms appeal particularly to parents traveling with children. They are also favored by groups of business men who wish to fold back the partition to make one large conference room so that they may "work on the way." With all these conveniences, each bedroom actually costs no more than two lower berths.

Connecting Double Bedrooms TYPE D

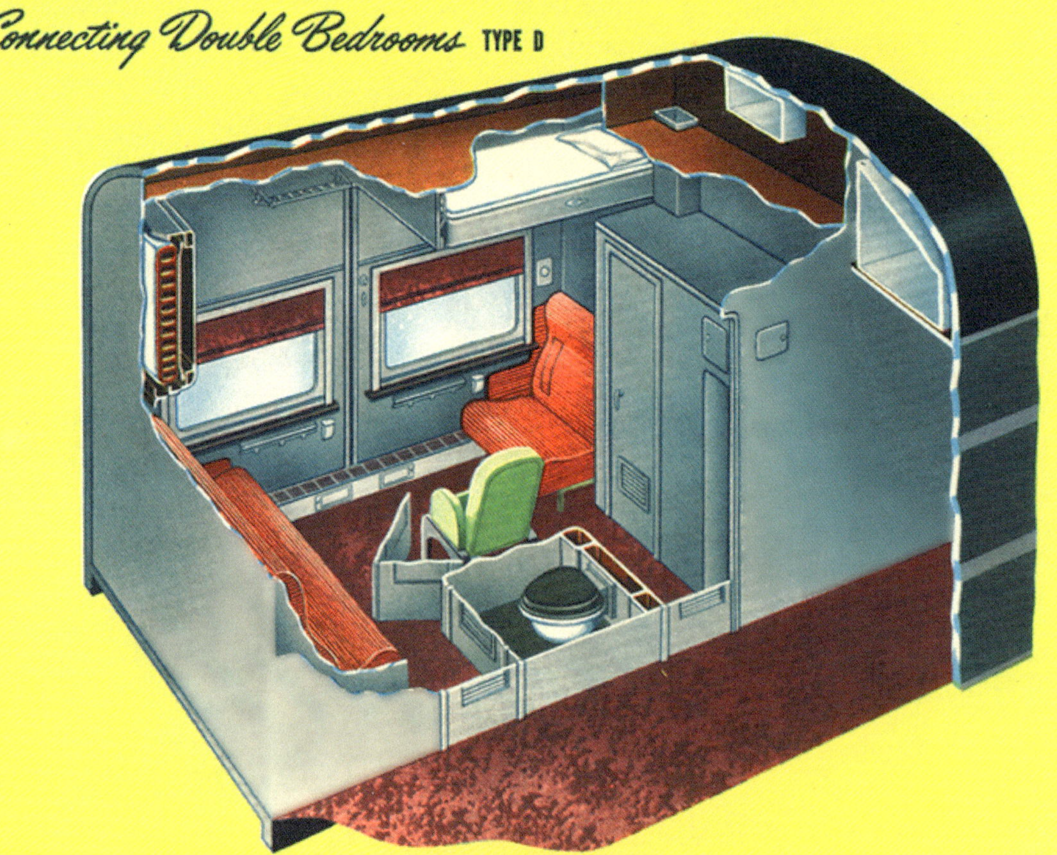

DAY

These two bedrooms, divided by a sliding partition, may be had singly or en suite. In one bedroom, which has a full-length sofa, the beds are *crosswise* to the car. In the other, which has a contoured seat and folding chair, the beds are *lengthwise.* Both rooms have full-size lower and upper beds. And each room is provided with enclosed toilet and washbasin facilities. Both rooms are air-conditioned and have separate heat control. They are built to accommodate a service table for games or writing. Lights are placed for both general and individual illumination, including mirror lights. Each room has circulating, mechan-

Connecting Double Bedrooms TYPE D

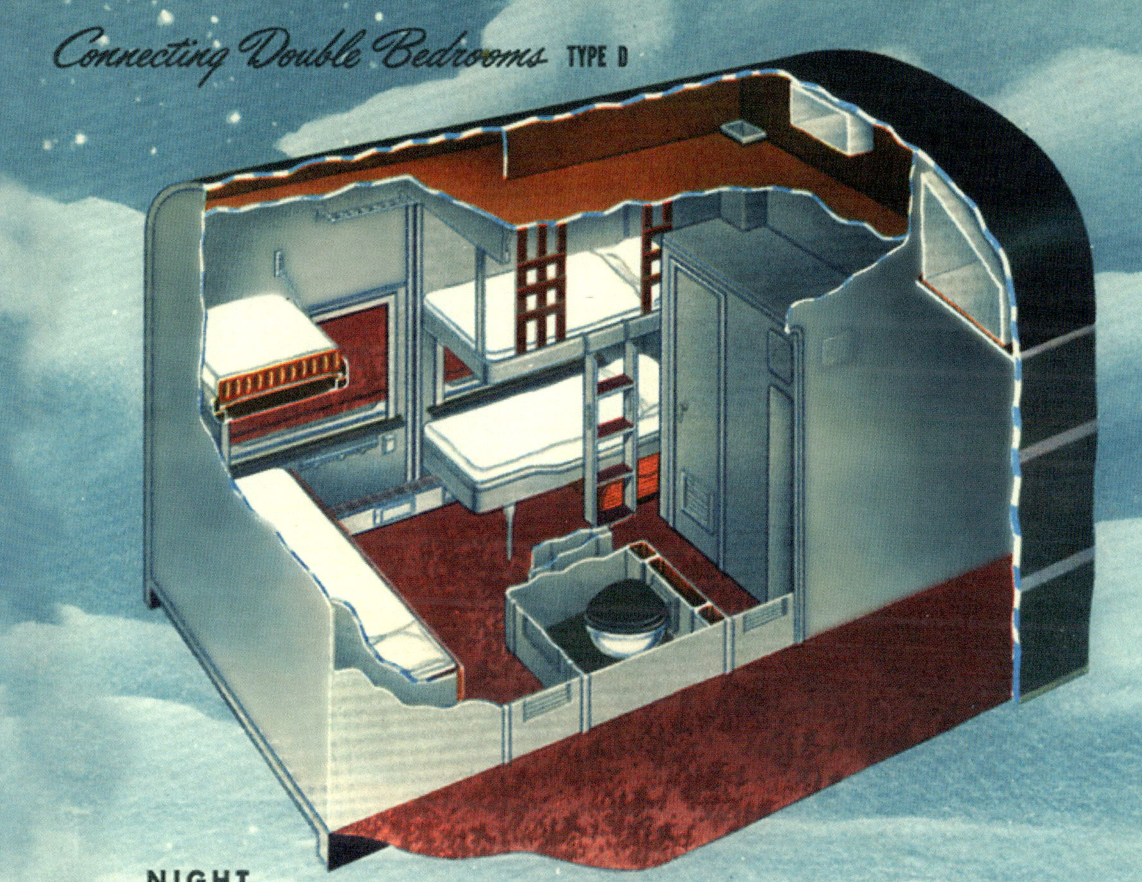

NIGHT

...ically cooled water. Other conveniences include a wardrobe locker, full-length mirror, electric razor outlets, ample luggage space, folding arm-rests, and shoe lockers. Each of these bedrooms accommodates two people. And the two together, with partition folded back, make an excellent accommodation for four. These bedrooms are favorites with families, and with business men who wish to make a "conference room" for several people. Another feature that has found favor is the connecting door that is built into the sliding partition. This allows the occupants to enjoy either company or privacy, whichever they choose.

Bedroom and Compartment

Here is a combination of accommodations often favored by three or four people traveling together. Although these private rooms may be had individually, the sliding partition between Bedroom and Compartment is easily folded to form one large room. The Bedroom, which has a full-length sofa that converts into a bed at night, has an upper berth above the sofa. It also has a stationary washstand and toilet that folds into a cabinet. The Compartment, which is lengthwise to the car, has a contoured seat and folding chair. It also has a full-size, folding type bed with an upper berth above it. Toilet and washbasin facilities are enclosed for privacy. Both

Bedroom and Compartment

NIGHT

rooms are air-conditioned and have separate heat control to suit individual preference. They also have lights for general lighting, individual reading, and mirror. Each has circulating water that is mechanically cooled, a full-length mirror, electric razor outlets, wardrobe locker, folding arm-rests, luggage rack, and shoe locker. As in all bedrooms, there is provision for a serving table. A connecting door in the folding partition enables the occupants to have company or privacy according to their preference. These combined accommodations are convenient for parents traveling with children regardless of age.

Compartment

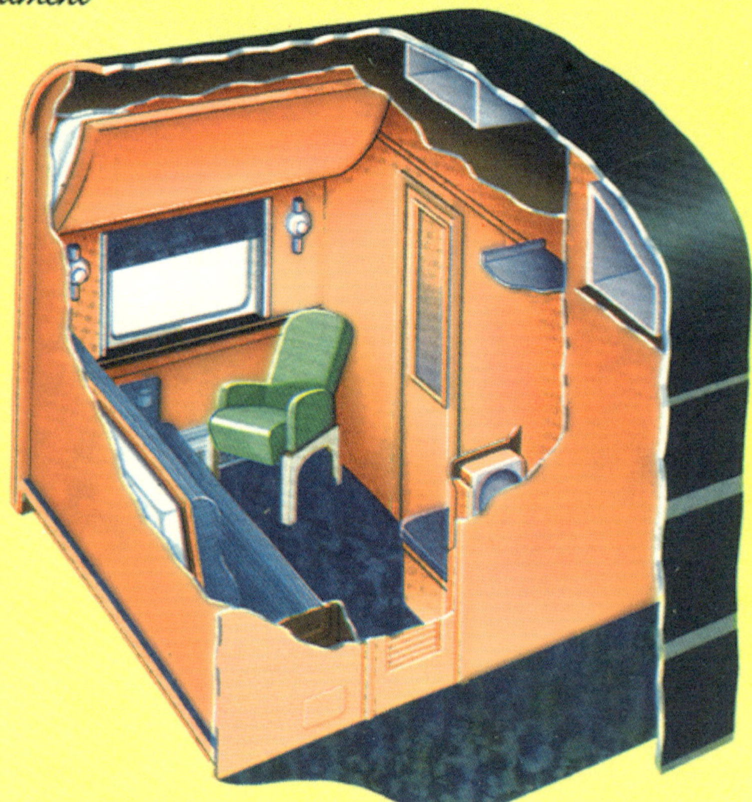

DAY

This accommodation is another example of obtaining additional space through rearrangement. Gone are the section seats by the windows, replaced by a transverse sofa affording ample lounging space. The upper berth remains above the window, and is at right angles to the convertible sofa-bed. There is space between the sofa and the opposite wall for a comfortable lounge chair. At night the chair is placed beneath the upper berth where it is still available for lounging and reading. During the day occupants have a comfortable sofa on which they may sit or recline, and also an easy chair. Upon request, the porter will set up a table

Compartment

NIGHT

for work, play, or food service. The compartment is fully air-conditioned and contains fan, individual regulation of lighting and ventilation, and thermostatic heat control. Other conveniences include folding arm rests in sofa back, call bells to summon porter at any hour, ash receptacles and mirrors. Each room has ample luggage space, complete toilet facilities, a locker for clothes, a shoebox and the latest designs in lighting fixtures. For daytime travel there is ample room to seat four passengers comfortably. At night there is a comfortable bed and upper berth, affording the finest in sleeping accommodations.

Drawing Room

DAY

Spaciousness is the impression received by travelers entering the Drawing Room. The enlargement has been attained by rearrangement of facilities, such as absorbing into the room the space previously occupied by the toilet and the entrance lobby, and providing enclosed toilet facilities. A longitudinal bed folding into one cross wall has been substituted for the former fixed couch; the section seats by the windows have been removed, with substitution of a long transverse sofa convertible into an equally long bed (with an upper berth) along one of the cross walls of the room. There is ample dressing space and easy access

NIGHT

to the enclosed toilet. For day travel the wall bed disappears, the upper is put away and the third bed becomes a sofa. This leaves space for two comfortable, movable lounge chairs, giving a real living room effect. The chairs are of the folding type, and at night are placed under the bed. The New Drawing Room has a fan, individual regulation of lighting, ventilation, air-conditioning, and thermostatic control of the heating. There is a large wardrobe; storage space above the enclosed toilet for luggage, with additional pieces going under the sofa, and a convenient box on the passage way in which to place shoes to be shined.

Observation Car

Here is one of the most cheerful rooms you can imagine. Extra-wide, clear windows at sides and back practically lure the sunlight into the car! All the comforts of a living room while traveling. Thick, soft carpets, deep, roomy chairs, handy tables and lamps all combine to give you a feeling of luxurious comfort. There is also a writing desk and accessories, current magazines, and smoking-stands for your convenience. The buzzer near your chair will bring an attendant to satisfy your wish for any additional service. Whenever you go Pullman, and there is an observation car attached, you are cordially invited to use it.

Lounge Car

Here is your private club on wheels. Deep, comfortable chairs and sofas invite you to relax, in the most pleasant kind of surroundings. From the thick, luxurious carpet to the tastefully appointed furnishings, everything here is designed for your comfort and convenience. You may enjoy the scenery, read the latest magazines, or enjoy the pleasant companionship of others. Convenient tables, reading lamps, and plenty of ash trays are all part of the pattern of pleasure in the lounge car. And there are desks and stationery if you wish to write en route. You're invited to make full use of the lounge car — it's there for you to enjoy.

Women's Dressing Room

This room is luxuriously equipped to help make a woman look her best. It's as spacious as a room at home, and fully as comfortable, with modern conveniences that include plate glass mirrors along one wall, a long dressing table, three well-upholstered dressing-table chairs, and a full-length mirror on the door. There are also plenty of wash basins, including one especially for brushing the teeth, separate toilet compartment, racks for bags, hooks for clothes, soap dispensers, and cosmetic tissues, with a generous supply of clean towels. And there's a wealth of well-diffused light. Here is every nicety of gracious living.

Men's Dressing Room

Here is a thoroughly masculine room a man can appreciate. In this commodious dressing room are all the conveniences anyone could want for shaving and dressing, including three or more large washbowls with well-lighted mirrors, a special basin for brushing the teeth, automatic soap dispensers, electric razor outlets, slotted containers for old razor blades, wall hooks for clothes, and a generous supply of clean towels. There's a wide leather sofa along one end of the room, and a smaller leather settee at the side. Above the sofa is a long, wide mirror — and a full-length mirror graces the door leading to the private toilet.

Like companionship? Ease back to the lounge car reserved for all Pullman passengers.

Want to "shake" the youngsters for a while? On some trains the railroads provide a nurse-stewardess who'll keep 'em happy.

Like to relax and stretch your legs? There's plenty of room on a Pullman.

all the conveniences . . .

Want a ladder for your upper? The porter brings it "pronto" when you ring.

Are you an epicure? As a Pullman passenger you'll appreciate the fine food served in the railroad dining cars that go with most trains.

Want a different view? Visit that lounge or observation car that goes with many trains.

Are you the "easy-going type"? Then *enjoy* the spacious comfort of the lounge or observation car.

Want to powder your nose? Each Pullman section car has a Women's Dressing Room where you can primp and preen in comfort.

Want to shave or change your shirt? The Men's Dressing Room is at one end of each Pullman section car.

your Pullman ticket +

Forget to write that letter? Dash it off at the writing desk in the lounge or observation car reserved for Pullman passengers.

Highball, cocktail, soft drink? Enjoy whatever suits your pleasure in the comfortable bar car that accompanies many trains.

Want to play cards? The porter will gladly bring a table that fits any accommodation you have.

Although we have emphasized the things you'll find *inside* your Pullman car, there is even more to be seen from your Pullman *window.* Here are the rugged hills and fertile valleys, mountains and plains, villages and farms of America—brought close and intimate through the magic of Pullman travel. Inside and out, there is a constant parade of pleasant contributions to your comfort and peace of mind when you

. . . GO PULLMAN

Your Pullman Ticket *Plus . . .*

Most trains that include Pullman cars also include dining cars, or other eating facilities, provided by the railroad.

Some trains include a bar car in which all kinds of mixed drinks and other refreshments are available.

On many trains you can purchase cigarettes, candy, playing cards, and other items.

On some trains you may have your clothes pressed if you desire.

There's even a barber at your service on a few trains.

On some trains the railroads provide an experienced nurse-stewardess to care for your children.

As you see, no effort has been spared to provide you with every possible comfort and convenience that will help make your trip a pleasure.

All for Your

SAFETY AND COMFORT

In these pages, we have given you a brief description of the many conveniences and comforts you can expect whenever you go Pullman. But these are just part of the picture.

Behind the scenes are the thousands of Pullman employes who constantly inspect the cars, keep them clean, and maintain the equipment that makes Pullman the *safest*, most *comfortable* way of going places fast.

And the courteous service you always receive when you go Pullman has to be experienced to be appreciated.

It is this constant concern with *your safety* and *your comfort* that has maintained Pullman's outstanding reputation for travel hospitality for more than eighty years.

Made in thc USA
Lexington, KY
20 July 2015